DATE DUE

APR 2 0 1999			
APR 1 0 2001			
APR 8 2003			
MAR 2 2 2008			
GAYLORD			PRINTED IN U.S.A.

Amelia Earhart

Oxford University Press, 198 Madison Avenue, New York, New York 10016

Oxford   New York
Athens   Auckland   Bangkok   Bogotá   Bombay
Buenos Aires   Calcutta   Cape Town   Dar es Salaam
Delhi   Florence   Hong Kong   Istanbul   Karachi
Kuala Lumpur   Madras   Madrid   Melbourne
Mexico City   Nairobi   Paris   Singapore
Taipei   Tokyo   Toronto   Warsaw

and associated companies in
Berlin   Ibadan

Oxford is a trademark of Oxford University Press

Text © Andrew Langley 1997
Illustrations © Oxford University Press 1997

Originally published by Oxford University Press UK in 1997.

Library of Congress Cataloging-in-Publication Data
Langley, Andrew.
    Amelia Earhart / Andrew Langley : illustrated by Alan Marks.
        p.   cm. – (What's their story?)
    Includes index.
        1. Earhart, Amelia, 1897–1937 – Juvenile literature.   2. Air
pilots – United States – Biography – Juvenile literature.
[1. Earhart, Amelia, 1897–1937.   2. Air pilots.   3. Women –
Biography.]   I. Marks, Alan, 1957–   ill.   II. Title.
III. Series.
TL540.E3L364   1997
629 .13'092--dc21
[B]
                                                                97–27400
                                                                CIP
                                                                AC
ISBN 0-19-521403-X

1   3   5   7   9   10   8   6   4   2

Printed in Dubai by Oriental Press

✦ WHAT'S THEIR STORY? ✦

Amelia Earhart

THE PIONEERING PILOT

ANDREW LANGLEY

Illustrated by Alan Marks

OXFORD UNIVERSITY PRESS

When Amelia Earhart was young, women were not expected to do adventurous things. But Amelia wanted excitement. She became the most famous woman flier in the world. Her courage and determination have inspired women ever since.

Amelia was born in 1897 in Kansas. Even as a little girl she did things her own way. She liked to jump over fences and gates. One day, her grandmother caught her. "Ladies don't climb fences, child," she said sternly. "Only boys do that."

Amelia did not see why girls should not climb fences. She enjoyed "boys' games," like baseball and roller skating. She liked to hunt rats in the barn with a rifle. Also, she always wanted to be the leader.

So, when she decided to build a roller coaster in her backyard, friends and family had to help. Amelia insisted on having the first ride—even though it ended in a crash!

Amelia liked to do things her own way. At school, she worked hard but refused to follow the crowd. Her schoolmates called her "the girl in brown who walks alone."

At Christmas 1917 she went to stay with her sister in Toronto, Canada. Here she had a shock. The First World War was raging in Europe, and many Canadian soldiers had been badly wounded. One day Amelia saw four men, each with a leg cut off, hobbling down the street on crutches. It upset her so much that she decided at once to give up college and work in a hospital.

For nearly a year she scrubbed floors, served meals, gave out medicine, and cared for the wounded servicemen. But she still found time to enjoy herself. She often went to the local airfield to watch the Canadian airmen flying. The sight thrilled her, and filled her with a great new ambition.

One morning, Amelia announced to her amazed parents that she wanted to learn to fly. She soon persuaded her father to pay for lessons, though he insisted on one thing: that she be taught by a woman. There were very few female pilots in those days, but Amelia found one. Her name was Neta Snook.

Early in 1921, Amelia arrived at the bumpy airfield for her first lesson in Neta's tiny biplane. Within a few weeks she was able to take it up into the air. She was flying!

Later that year, she had her first crash, and ended up in a cabbage field. She told her friends the accident had put her off cabbages, but not off flying.

Now she wanted to buy her own aircraft. She set her heart on a little bright yellow plane, nicknamed the Canary. Taking part-time jobs, she began to save. Just over a year later, she was the proud owner of the beautiful little aircraft.

Nearly a year after her first lesson, Amelia took one more big step: She flew solo. To celebrate her safe landing, she bought herself a leather flying coat. An even better present was her pilot's license, which she was given in 1923. She was only the 16th woman in the world to have one.

Flying was the most important thing in Amelia's life, but she still had to earn a living. She took a job in Boston teaching foreign children, mostly Chinese, who had come to live in the United States. She loved looking after the young girls and boys. And they loved her—especially when she drove them in her bright new sports car, the Yellow Peril.

One day in 1928 Amelia was summoned from her classroom. There was a phone call for her. She picked up the phone and heard an amazing question. Would she like to be the first woman to fly across the Atlantic Ocean?

The year before, Charles Lindbergh had flown solo across the Atlantic. He had become an overnight hero. Now George Putnam, a rich man, was planning the first flight across the ocean by a woman. He said he wanted "the right sort of girl" to make the flight.

Amelia was just right. She even looked like Charles Lindbergh! She was thrilled to have been chosen. And so she found herself, one windy morning, climbing aboard the seaplane *Friendship* on the Newfoundland coast of Canada. She was traveling as a passenger—the pilot and navigator were both men.

Amelia's excitement grew as *Friendship* lifted off the waves. The plane climbed slowly up into the sky. Soon there was thick fog all around. Amelia made notes about the height and speed. After 20 long hours, the pilot spotted a huge bay below and made a perfect landing. When Amelia wearily opened the hatch, she saw a small boat alongside. "We've come from America," she called. "Where are we?"

*F*riendship had landed on the coast of Wales. The next day, the three of them flew on to Southampton. A huge crowd was there, all wanting to see Amelia. She was the star—all because she was a woman. Her journey was a sign that women could be adventurous.

Amelia said that she was "just baggage," but it made no difference. The rest of the crew were ignored. And when they got back to the United States, Amelia was the nation's heroine. With the pilot and navigator, she paraded through the streets of New York.

Amelia knew she still had to prove that she was a good pilot. Encouraged by George Putnam, she took part in the first all-women air race across the United States, and came in third.

During the race, she became friends with many of the female fliers. They all thought that women should have the same opportunities as men. They formed a society just for women pilots. It was called "The Ninety-nines," after the number of women who applied to join.

Now that she was famous, Amelia could be a full-time pilot. She earned money by appearing at air shows and writing books about her adventures. She also began to set new flying records. During 1930 she flew faster than any woman had ever done before—not once, but three times!

The following year Amelia married George Putnam. He managed her career, organizing flights and public appearances. He sold her stories to newspapers and dreamed up stunts to make her even more rich and famous.

One of these stunts involved an early kind of helicopter called an autogiro. It was nicknamed "the flying windmill." In this, Amelia set a new height record for both men and women. Then she flew it right across the United States in nine days. But the autogiro was difficult to control, and she crashed twice on the way. Luckily, she was unhurt.

"Would you mind if I flew across the Atlantic?" Amelia asked her husband one day.

It was early 1932. Even now, no woman had ever made a solo flight across the ocean. It was a dangerous journey. Many people thought it was impossible. But Amelia saw her chance to make a real mark for female pilots.

She took off from Newfoundland. A lightning storm began as she flew higher. After several hours the plane seemed to grow heavier. It was being covered in ice from the clouds. Suddenly, the plane went into a spin. Down it plunged, until Amelia could see the whitecaps of the waves below. But as she went lower the ice melted, and at last she could climb again.

After nearly 15 hours Amelia saw land ahead. She circled around and came down safely in a field of cows. An amazed farmhand ran up to tell her that she had reached Ireland.

Back home in New York there was a massive welcome for Amelia. Ships hooted and whistled. Navy planes swooped overhead. Thousands of cheering people watched her drive by, through a blizzard of torn-up paper and ticker tape.

She was now an even bigger hero than ever. She was awarded a medal for her flight and dined with film stars, sportsmen, and presidents. She started a tour of the West Coast, giving lectures about her adventures. She even became a fashion designer, producing a collection of women's clothes.

But Amelia did not enjoy the publicity and the traveling and the crowds of fans. She was much happier up in the air and away from other people. That summer, she flew nonstop across the United States. The flight took 19 hours and covered 2,418 miles (3,900 kilometers), longer than her Atlantic crossing. It was yet another record.

Amelia had flown across the Atlantic. Now she looked for a new challenge. She decided to fly across the Pacific Ocean—or at least part of it. She planned to cross from the Hawaiian Islands to California. That way she would not get lost: "It's easier to hit a continent than an island!" she said.

Amelia traveled out to Honolulu by boat, with her aircraft lashed to the deck. She had to wait for many days before the weather was right. At last, she was able to squelch down the muddy runway and take off.

It was a horrible flight. Thick fog and clouds surrounded her all the way, so that she did not know exactly where she was. A cover blew off, letting a stream of icy air into the cabin. The freezing air made one eye swell up. Then the fuel began to run short, so she had to slow down. Amelia was very glad to land at Oakland, two hours later than planned.

Amelia was tired out with her lecturing and traveling. It was not surprising that she got sick and had to rest in a hospital. But that did not stop her from making new plans. Her next flight would be the biggest one of all: around the world, along the equator. No one had ever tried this before.

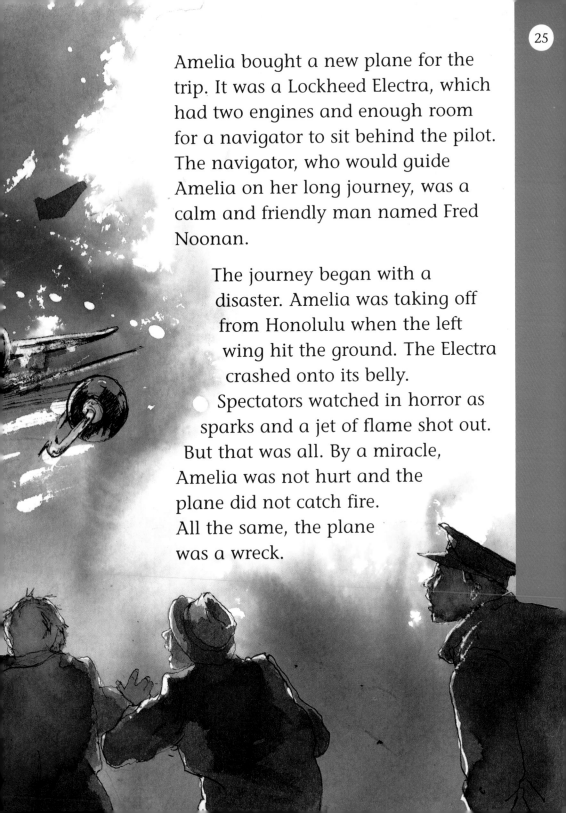

Amelia bought a new plane for the trip. It was a Lockheed Electra, which had two engines and enough room for a navigator to sit behind the pilot. The navigator, who would guide Amelia on her long journey, was a calm and friendly man named Fred Noonan.

The journey began with a disaster. Amelia was taking off from Honolulu when the left wing hit the ground. The Electra crashed onto its belly. Spectators watched in horror as sparks and a jet of flame shot out. But that was all. By a miracle, Amelia was not hurt and the plane did not catch fire. All the same, the plane was a wreck.

PACIFIC OCEAN

New Guinea

It took nearly two months to repair the Electra. During the wait, Amelia changed her plans. She now decided to fly around the world eastward instead of westward.

At last the aircraft was ready. On June 1, 1937, Amelia and Fred Noonan took off from Miami on the first leg of their great adventure. They flew along the coast of South America to Brazil, and then across the Atlantic to Africa and Asia.

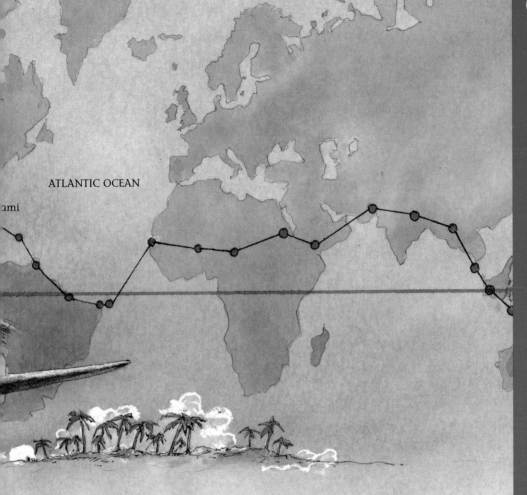

ATLANTIC OCEAN

ami

Each day, Amelia was at the controls for at least 10 hours. Each evening, they would land and sleep for about five hours. Then they were up before dawn to take off again. The Electra flew on and on, over jungles and deserts and mountains and seas. After three weeks they reached New Guinea, on the edge of the vast Pacific Ocean.

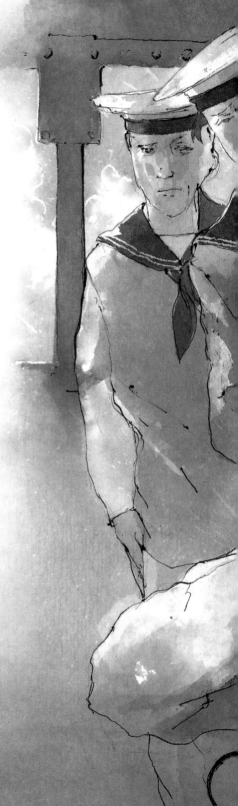

Now came the most dangerous part of the journey. In all that huge empty space of sea, Amelia had to find one tiny speck of land: Howland Island. She had to land there for more fuel. It would be very easy to miss.

On the morning of July 2, Amelia set out from New Guinea. People watching saw the Electra roar off the end of the runway and then drop away toward the sea below. But it had just enough speed to climb above the waves and up to safety. After a few minutes, it disappeared from view.

This was the last sight anyone had of Amelia, or Fred, or their aircraft.

Several times during that day, Amelia's voice came over the radio. Everything seemed to be all right. Hours passed. The radio calls became less regular. Faintly, Amelia could be heard saying that fuel was low and they could not see land. Then the calls stopped altogether. Amelia Earhart had vanished.

Where was the Lockheed Electra? Amelia must have missed the island and come down in the sea somewhere nearby. So, during the next two weeks, 10 ships and 65 aircraft made a desperate search of the area. They found nothing.

Since then, no trace has ever been found of Amelia. And nobody knows exactly what happened. Some people said that she ran out of fuel and crashed. Some said that she floated for days in her life raft. Some said she was eaten by sharks. Some even thought that she had been kidnapped by the Japanese!

One thing is certain. Amelia Earhart is still the most famous of all female fliers. Daring and determined, she showed that women could follow their own dreams in life—no matter how dangerous they were.

Important dates in Amelia Earhart's life

1897 Born in Atchison, Kansas.
1916 Graduates from high school in Chicago.
1917 Works in a hospital in Canada for the Red Cross service.
1919 Becomes a student at Columbia University.
1921 Begins flying lessons.
1922 Buys first airplane: a Kinner Canary.
1923 Receives pilot's license.
1926 Becomes a teacher in Boston.
1928 Flies the Atlantic aboard the seaplane *Friendship*.

1929 Flies in the first women's air race; founder member of "The Ninety-nines."
1931 Marries George Putnam; sets new height record in autogiro.
1932 Becomes the first woman to fly solo across the Atlantic Ocean.
1935 Makes the first solo flight from Honolulu, Hawaii, to California.
1938 Disappears in the Pacific Ocean on around-the-world flight.

Index